WHAT MAKES YOU UNIQUE?

The Secrets of Genes and Heredity

WHAT MAKES YOU UNIQUE?
The Secrets of Genes and Heredity

Lin He
Chinese Academy of Sciences, China

Bo Hai
Chang Qin
Shanghai Media Group, China

Translated by

Yushu Wang
Shanghai Jiao Tong University, China

 **World Scientific**

NEW JERSEY · LONDON · SINGAPORE · BEIJING · SHANGHAI · HONG KONG · TAIPEI · CHENNAI · TOKYO

Published by

World Scientific Publishing Co. Pte. Ltd.

5 Toh Tuck Link, Singapore 596224

USA office: 27 Warren Street, Suite 401-402, Hackensack, NJ 07601

UK office: 57 Shelton Street, Covent Garden, London WC2H 9HE

Library of Congress Control Number: 2021931311

British Library Cataloguing-in-Publication Data
A catalogue record for this book is available from the British Library.

《基因要去串门了：基因和遗传的秘密》
Originally published in Chinese by East China Normal University Press Ltd.
Copyright © 2017 by East China Normal University Press Ltd.
English translation rights arranged with East China Normal University Press Ltd.

WHAT MAKES YOU UNIQUE?
The Secrets of Genes and Heredity

Copyright © 2021 by World Scientific Publishing Co. Pte. Ltd.

ISBN 978-981-123-248-0 (hardcover)
ISBN 978-981-123-249-7 (ebook for institutions)
ISBN 978-981-123-250-3 (ebook for individuals)

For any available supplementary material, please visit
https://www.worldscientific.com/worldscibooks/10.1142/12166#t=suppl

Desk Editor: Ling Xiao

Printed in Singapore

Preface

The launch of "Talk on the Sea" studio is a dream for us as broadcasters. Information and communication technology changes with each passing day, and changes to the way of communication brought about by new technologies have impacted the traditional media such as newspapers, periodicals, radio and television beyond our imagination. With the rise of Internet technology and mobile reading habits, several newspapers have announced their closure and most traditional media are seeking transformation. Will traditional media die? This is a question of many newsmen.

As a historical, ancient and traditional form of media, radio has not disappeared under the impact of Internet technology but has sprung up with vitality despite the turbulence, which is beyond expectation but is also reasonable. Today, the colorful radio programs are inseparable from the broadcasters who have been striving tirelessly for years. Under the constant impact of new technology, the broadcasters always grasp the information content, to drive innovation in the program using new technology, to actively seek innovation and change, and to find more opportunities during changes in technology.

In the new era, what should media professionals do with the strategic demands of "building a center for technological innovation with global influence"? How can they do to contribute to creating the social atmosphere of "mass entrepreneurship and innovation"? Can the media achieve "self-innovation" in the form and the content of communication methods and means? How can we let the concept of supporting innovation and tolerating failure be readily accepted? The "Talk on the Sea" program tries to answer these questions.

We have exclusively planned the radio series "Questions About Innovation — Dialogue Between Elementary School Students and Chinese Academicians", in an attempt to build fertile soil for fostering innovation. The original intention of this program is to invite Chinese academicians to talk with elementary school students about the current interesting popular science topics. We believe that children at the elementary level have strong curiosity and a thirst for knowledge. They have strange ideas and their questions are unique and tricky. Will conversations between academicians, who are experts in their field, and elementary school students, who speak the truth, bring surprises?

We invited Chinese academicians to bring popular science to elementary school students. Things went unexpectedly smoothly. The academicians expressed their support, and the program was successfully completed. The content of the talks for the students was not particularly cutting-edge science, but more towards foundational science and scientific exploration for technological breakthroughs, which in turn drives industrial upgrading and ultimately serves mankind. A journey of a thousand miles begins with a single step. The road to scientific exploration is long and difficult. The academicians shared their growing-up experiences in the form of stories about their own "learning" with the children, and guided them to develop the habit of "thinking".

The scientific knowledge given by the academicians for the

children was not only theoretical research content, but also knowledge combined with China's current industrial status. The purpose was to enable the children to experience the current industrial situation, to understand the background knowledge of professional disciplines, to create their awareness of the profession, and to make them realize that the dream of a technologically advanced country must be based on reality.

The nonagenarian Shuhua Ye, a well-known astronomy expert and an academician, announced for the first time, on behalf of the scientific community, China's participation in the Giant Telescope Project of the world's space exploration program. "Talk on the Sea" was the first program to disclose the news in the country. The academician Xuhong Qian told the story of how he dismantled the alarm clock when he was a child so that the children could have a better understanding of working with the brain and the hands. The serious-looking face of the academician, Shichang Zou, allowed the children to feel the aura of the elderly scientist. The atmosphere in which the academician, Lin He, talked about genetics was very lively, and his discussion about why twins look alike delighted the audience. The conversations were filled with wit and wisdom, and everyone was in for an auditory feast! The academicians did not stick to the traditional instillation of popular science knowledge, but rather displayed their personal charm, which drew the students closer to them. During the talk, the children shot bold and unconstrained questions at the academicians who handled their questions patiently and even replied frankly with "I don't know" to encourage them to think and explore on their own. The listeners were not only surprised by the knowledge of the elementary school students, but also touched by the care the academicians had for the precious exploratory spirit of each child.

Not only had elementary school students a hunger for

knowledge of popular science, but also junior high school students.

We could not forget the lively dialogues even after the program ended. We hoped that more children could listen to the talks by the academicians. Thus the book series on "Dialogue with Chinese Academicians" was produced. With the support of the participating academicians, we systematically expanded the content knowledge of the talks in the form of text with illustrations to present the basic knowledge of the academic subject clearly and vividly. A group of young PhD students striving in various scientific research fields joined the writing team to compile the content knowledge. They organized the knowledge of different areas from the talks and supplemented the relevant professional content to make the book series scientifically more three-dimensional and intellectually more fulfilling.

The academician Xiongli Yang said during the program that science is about dealing with new things and continuous innovation. We dedicate this series of books to the students, hoping that they can have fun exploring one secret after another in their growing up years.

Bo Hai, Chang Qin
"Talk on the Sea" program

February 26, 2017

Contents

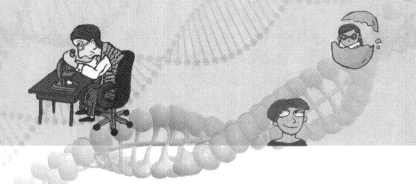

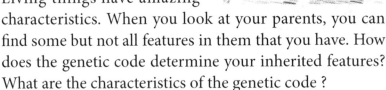

Chapter 4

Can Genes Communicate with the Universe?

You are a part of the earth and also a part of the universe, and you are integrated into the universe on this earth. The human body is a closed open system that is constantly communicating with its surrounding environment, forming a communication between human genes and the universe.

Chapter 1

No shortcut to Success

" There is no short cut to success. The geneticist Professor Lin He tells you why by sharing stories from his growing-up years. Professor He does not set goals but needs to see the direction. He then walks in the right direction one step at a time and harvests fruits of different sizes along the way. "

Bo Hai: Hello, I'm Bo Hai. Our series "Dialogue with Chinese Academicians" is specially prepared for you. We have invited great scientists who are striving in the scientific research fields to share their growing-up stories with you and the scientific knowledge that you want to know. The stories of these great scientists are both interesting and inspiring, which will motivate you to make your resolutions early, and maybe someone here will become a great scientist in the future.

Chang Qin: I'm Chang Qin. With me here is Professor Lin He and he will talk with you today. Do you know what kind of research he does?

Students: Yes, we know! He is a Chinese academician and a genetic biologist who is an expert in gene research.

Professor Lin He

Lin He is an academician of the Chinese Academy of Sciences and a well-known genetic biologist, who has led the establishment of one of the world's largest sample libraries for mental and neurologic diseases. Once, Professor He drove alone to Gansu province in order to collect some samples. He was fatigued and nearly caused his car to tumble down the mountain. It was such bravery that led Professor He to finally solve a century-old genetics puzzle and discover a genetic disorder, which was named after him, living up to his name as the genetic code translator.

Lin He: I'm very moved to see so many young friends today. I have been thinking: What was I doing at your age? I would like to share my story with you.

I was an innocent and a very happy child. A problem with my health, however, caused me great sadness. When I was at your age, I was prone to asthma attacks. Each time an attack happened, I had to go to the hospital for treatment and could not go to school. I fell behind with my schoolwork as a result. Fortunately, my school grades were average. But I could not go to secondary school after graduating from elementary school because the "Cultural Revolution" had begun.

科学之问

Chang Qin: There was no need to go to school then, right?

Lin He: Yes, I was mostly either at home or outside wandering during that period.

Bo Hai: How can a child who was often sick at elementary school and wandered during secondary school years finally become an academician of the Chinese Academy of Sciences? I'm really curious!

I could not go to school at that time. After staying home and wandering outside for several years, I came across an opportunity to resume studying. From 1969 to 1970, the opportunity to resume classes for more than six months came in the form of learning industrial skills and farming, which means learning the techniques used in factories and learning to farm. I not only learned weaving techniques, but also learned how to operate the walking tractor and most of the field work. After passing through the "secondary school" period in a hurry, I was sent to a factory and stayed there for 8 years. I missed almost the entire secondary school learning opportunity. How can I compete with those high school students to enter the university? How shall I deal with the university entrance examination? Looking back on those days, I still feel somewhat scared and intrigued. At that time, I had a friend who was also my table tennis mate. He did not graduate from university because of the "Cultural Revolution" and was assigned to work at my factory. Because he loved playing table tennis as much as I did, we often played together and built up our friendship. Then, Xiaoping Deng came into power and gave everyone the opportunity to take the national college entrance examination. It was the first college entrance examination after almost 12 years, the competition of which was the strongest in the history because of the many applicants. With the help of my friend, my dream finally came true as I stood out from the competition in 1978.

My friend helped me a lot with his accumulated solid foundation from secondary school. Having missed secondary schooling, I had to learn from the beginning rather than review systematically what I knew. What should I do under the grim circumstance? The trick was to predict questions! My friend helped me to predict questions like playing a betting game. "You have to think of a way to remember this question, and also for that one," he said. In fact, I learned everything without thinking to make up for my lack of knowledge. It was impossible for me to really understand and remember at that time.

At that time, I could only learn by rote, but no matter how hard I tried I could hardly memorize. How could anyone memorize the contents of mathematics, chemistry and physics? Therefore, I simply could not understand many questions. So, do not use this method to learn. The result turned out to be not bad in my opinion, and rote learning did help me get a taste of real knowledge. How many of you can know how I feel? I had to endure working three shifts and high-intensity study. By the time I passed the ordeal to secure a place in college, I was already 25 years old.

Chang Qin: You entered college at the age of 25.

Lin He: Yes, eight years of working at the factory was a huge waste of time, but I also gained a lot of knowledge that cannot be learned in school.

Chang Qin: After listening to Professor He's story, will the elementary school students here think like this: "I don't have to study much in elementary and secondary school. Like Professor He, I will find a good friend who can tutor me, then I can go straight to university by passing the college entrance examination"?

You seem to be quite interested in the history of my college entrance examination, so I will talk a little bit more.

When I look back on my experience of taking the college entrance examination, it seems as easy as dreaming, but also as difficult as crossing the glaciers. Because of missed schooling and the tough working conditions, I had to think of a plan to be prepared for the college entrance examination. My plan then was to seek wisdom and diligence. I was not able to sleep during my night shift, and felt like I was walking on cotton, with my head feeling heavy. Even then, studying really hard to prepare for the exam is a must. The competition is like crossing a single-log bridge, on which taking every step involves a risk. When I calculated the total time I spent on the preparation, to be able to enter college seemed like a joke. Considering the efforts I had put in, the process seemed like "peeling a layer of skin off my body", but the outcome was not bad.

Therefore, there is actually no shortcut to success. Even though talent is needed in the strive for success, it is not possible to achieve the ultimate result by seizing every opportunity. Of course, friendship plays a role in success.

"You do not have to set goals but need to see in the right direction, taking one step at a time wherever you are. Only in this manner will you achieve unexpected results, and not be frustrated by not meeting your goals."

— A message from
the academician, Lin He

Chapter 2

Looking for Similarities and Differences

" In the gene is found the amazing nature of heredity. Whom do you look like? What characteristics do you have? What special personality do you have? Are you able to find the answers by looking for similarities in your own family? "

Questions to Ponder

What does each person look like?
What characteristics does each person have?
How is the uniqueness of each person reflected?

Hint

When reading this chapter, you can think about these questions first. You may be able to answer them after you finish the chapter. If you are still interested in these questions after this chapter, you can also search online for the relevant knowledge.

Bo Hai: Do you know that Professor Lin He's father is also an academician?

Lin He: While I am on this topic, I'd let you know that my daughter has managed to be admitted to the University of Cambridge.

 Bo Hai: Can it be explained by the good "learning" genes in Professor He's family?

Lin He: It is possible that there is an "academic" gene at work. It is still difficult to conclude which type of genes are responsible though.

Genes Bringing Similarity

Genes possess special hereditary abilities. How do you look like? What characteristics or special personalities do you have? The answers may be found when you go and compare yourself with your parents. You will find quite some similarities.

You may have noticed that parents with big eyes will likely have a child with big eyes. Some of us are short-sighted, just like our parents. A photo of a mother taken as a child may look the same as her daughter of about the same age. Such observations are not surprising.

The word "genes" refers to the genetic factors in the human body that are DNA segments with genetic effect. When you were an embryo or still in your mother's belly, your physical appearances, like the type of nose, mouth, hair, fingers, and even nails, are developed under the guidance of genes. Such physical characteristics are what biologists call "traits".

The process of development is influenced by not only genes but also the external environment. Pregnant mothers therefore need supplementary nutrition to help the fetus grow better. After birth, you continue to grow and develop as instructed by your genes.

External environmental influences are also important. For example, you should have the genes to grow tall if both your parents are tall. But you may not grow too tall if you do not eat well for good nutrition. There are many examples of how acquired factors can affect the process of growth and development. Remember, if you want to grow tall or beautiful, do not be a picky eater but rather take in more nutrition, so that your genes can work well!

One may be slow tempered, like one's parent. Another may be as stubborn as the parent. You may be asking questions: Can personality traits be inherited? Is character influenced by genes? The formation of personality is related to the personal growth environment. Family members have a great influence on personality. It is difficult to conclude how big a role genes play in personality.

Student: I think my thumbprint is similar to my mom's but is not exactly the same as hers.

Lin He: Your expression is clear and precise.

Student: I read in a book that only two out of 60 billion people will have the exact same fingerprints.

Chang Qin: Is that true?

Uniqueness of Genes

The possibility of finding two people with the exact same fingerprints in this world is extremely low. In 1892, the British scientist, Francis Galton, stated in his book *Finger Prints* that the chance of one pair of fingerprints being identical is less than one in 64 billion people. However, this view has not been scientifically proven. There are now nearly eight billion people in the world. No two people with identical fingerprints have been found yet, proving the low possibility.

Fingerprints are unique. You may have watched on TV how the police identify a suspect using fingerprints found at a crime scene. Fingerprints are genetically unique and can be used for individual identification.

In biology, DNA fingerprints are used to identify individuals.

Using DNA to Solve Crimes

Some time ago, news came that the serial murder case in Baiyin, Gansu province of China, had been finally solved. It was reported that 11 people were murdered in a period of 14 years from 1988 to 2002. The cunning criminal left very few clues at the crime scene, and the police were unable to solve the case for many years. But the investigators never gave up their efforts. They carefully picked up the clues left by the criminal at the crime scene and studied them, focusing on scientific analysis of the collected biological evidence using new technology. Using the Y-chromosome DNA test, the police suspected the Gao family in Chenghe village. The police recorded the family's fingerprints one by one, compared the fingerprints with the DNA, and eventually identified the criminal. Genetic identification technology played a crucial role in the detection of this case.

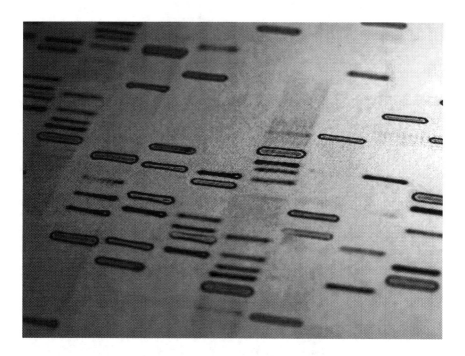

Just like the barcode of commercial products, DNA fingerprints can be presented by a series of stripes on a sequencing gel,
and they are highly specific for each individual. DNA fingerprints
are so named because they are comparable to the fingerprints of
human fingers for recognition. Unlike the fingerprints of human
fingers, DNA fingerprints cannot be changed or erased, and only
one drop of blood or one stand of hair is needed to identify a
person's DNA fingerprint. DNA fingerprints can thus be used in
identification of the individual and paternity testing.

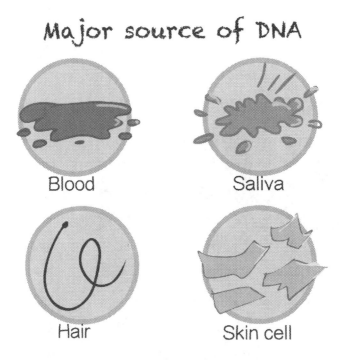

Major source of DNA

Blood   Saliva

Hair   Skin cell

Chang Qin: Which part of you resembles your dad or your mom?

Student: My eyes are like my mom's. Hers are small and so are mine.

Student: I look more like my mom. My eyes are very similar to hers, my nose is as sharp as hers, and my little finger also looks like hers.

Bo Hai: What do you mean that your little finger looks like your mum's?

Student: My dad's little finger has the same width throughout, while mine and my mom's are wide at the base and narrow at the tip.

Basics of Genes

Many similarities and differences exist between children and their parents. To understand the characteristics of heredity, you need to have some basic knowledge of genes.

Structure and Composition of DNA

Your genetic information is stored in the DNA. Let's get to know the DNA that makes up your genes. DNA is short for deoxyribonucleic acid and its basic unit is the deoxyribonucleotide. Each deoxyribonucleotide comprises a phosphoryl group, a five-carbon sugar and a base containing nitrogen (N).

phosphoryl group

sugar

DNA Replication

The DNA structure provides the basis of DNA replication. DNA replicates when a cell is about to divide into two new daughter cells. DNA replication is a semiconservative process. It requires separating the double stranded DNA into two strands. Each strand serves as a template for a new strand with the same substance.

Genetic Code in DNA

Your genetic information is stored in the DNA. The DNA is replicated to be passed on to the daughter cells. The DNA sequence is transcribed into RNA that guides the protein making process. These steps form the "central dogma" of genetics.

DNA → RNA → Protein

The central dogma of genetics was proposed by Francis Crick in his article "On protein synthesis" published in 1958. The central dogma states that genetic information is transferred from DNA to RNA, and then from RNA to protein; and genetic information is transferred from DNA to DNA by replication. With the development of molecular biology, the question is whether the central dogma will be accepted or challenged. The road to science is sometimes difficult and painful, but sometimes interesting and filled with imagination. It is only when you devote yourself in it that you will feel the joy.

DNA Mutation

When DNA replicates, errors may occur. One or more bases may be inserted or deleted in the DNA sequence, or a wrong base may be used in the sequence. These errors lead to changes in the DNA sequence and thus in the genetic information.

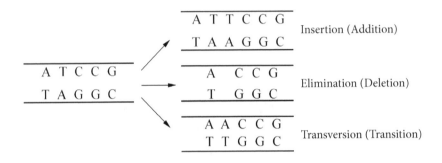

Discovery of DNA

The American scientist James Watson and British scientists Francis Crick and Maurice Wilkins shared the Nobel Prize in Physiology or Medicine in 1962 for their research on the double helix structure of DNA.

The proposal of the double helix structure of DNA was the most exciting moment in the history of genetics. Since the 1950s, the rapid development of molecular biology led to the study of genetics at the molecular level. The discovery of the DNA double helix opened the door of the "mystery of life" to a clear understanding of the composition and transmission of genetic information. It was also from the 1950s that disciplines such as molecular genetics, molecular immunology and cell biology flourished and paved the way for the application of science and new technologies.

The literature tends to present the discovery of the DNA double helix in the following way:

Rosalind Franklin, a British female scientist, went to France to study X-ray diffraction technology and achieved much in the field of physical chemistry. In 1951, Franklin was hired to work at King's College London in the department of the famous physicist John Randall. Randall assigned her to investigate the structure of DNA.

Franklin was not the only person in the study. Wilkins, who later won the Nobel Prize, had been in the team studying DNA for some time. Franklin's excellent X-ray crystallography technique won the respect of Wilkins, even though he did not like her involvement in the research. In May 1952, Franklin, with the help of Wilkins's student Raymond Gosling, took an X-ray diffraction photo of B-form DNA, known as "Photo 51".

Rosalind Franklin

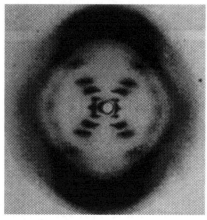

Photo 51

Physicists of that time analyzed the structure of crystals by looking at X-ray diffraction photos. When X-rays penetrate the crystal, a diffraction pattern of alternating light and dark bands forms. Crystals with different structures produce naturally different patterns. Scientists can speculate the atomic arrangement of the crystal based on the pattern. Franklin did not immediately publish the research result of the photo, which led to the later controversy around who discovered the double helix structure of DNA.

In addition to King's College London, the rival Cavendish Laboratory at Cambridge University also had a group of scientists studying DNA. At that time, the active academic atmosphere in London allowed many young and innovative scientists to gather often, to talk and express themselves freely, and to even hold debates criticizing the authority of that time. Watson and Crick were studying DNA in the Cavendish Laboratory, and they got to know their so-called competitor, Wilkins. How interesting!

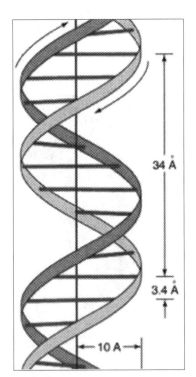

DNA double helix

Scientists at that time speculated that DNA might be the genetic material, but what its structure was and how it functioned in life were unclear. The triple-stranded helix structure for DNA was the hypothesis then, but no one knew the real DNA structure.

In January 1953, Watson saw Photo 51 when Wilkins showed

him. The photo immediately served as an inspiration. The DNA structure that Watson and Crick had tried to figure out for months suddenly became clear to them: two strands with the phosphoric acid forming the backbone are intertwined into a double helix structure, and connected together by hydrogen bonds. On April 25, 1953, Watson and Crick announced the discovery of the double helix structure of DNA molecule in the British journal *Nature*. The article was a milestone for biology and the beginning of the era of molecular biology. Its importance cannot be underestimated.

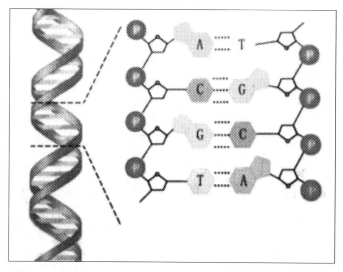

DNA structure

Franklin knew nothing about Wilkins showing her Photo 51 to Watson. Wilkins showed Watson the photo without Franklin's permission, and Watson and Crick also used the photo without her permission. Franklin later also published an article in the journal *Nature* deducing the double helix structure of DNA.

In 1962, when Watson, Crick and Wilkins shared the Nobel

Prize, Franklin had already died of illness. Franklin's name was left out from the history of the Nobel Prize.

Several years later, Watson and Crick both acknowledged that Franklin's finding was a vital clue to constructing the double helix structure. Crick mentioned in an article that Franklin's contribution had not been adequately recognized and that she had clearly explained the two types of DNA and determined the density, size and symmetry of the A-form DNA. In 2000, a new building at King's College London was named "Franklin–Wilkins building". At the official opening, Watson said: "Without the two of them (Franklin and Wilkins), Francis Crick and I could not have made our discovery."

This story is winding and filled with sadness. If Franklin had not died so early, would she have won the Nobel Prize?

James Watson giving a speech at Shanghai Jiao Tong University (hosted by Lin He)

James Watson visiting geneticist Jianzhen Tan (also known as C.C. Tan) at Huadong Hospital with Lin He

Further Thinking

After reading this chapter, you may think about these questions: Why did I become me? What makes me unique? You can start by reading the story about Dalton's description of color blindness. After reading it, think about the questions again and write down your thoughts.

John Dalton may be the first person to clearly describe his color blindness condition. He was not a biologist but a famous British chemist and physicist. He was the founder of modern atomic theory, and the atomic mass unit is named after him.

It was said that Dalton wanted to buy a pair of silk stockings as a present for his mother. He looked carefully and chose a dark, dull blue pair. He gave the present to his mother and said, "Mom, you'll be happy with this pair of stockings." Seeing the stockings, his mother said, "You have bought me a pair of stockings, John, but what made you choose a bright-colored pair?" The surprised Dalton replied, "This dark, dull blue pair suits you." His mother replied, "They're as red as a cherry, how can I wear them?" Puzzled, Dalton asked his younger brother, who agreed with him. But when their neighbors were asked, all said that the stockings were red.

Dalton did not let this matter rest. After careful analysis and comparison, he found that he and his brother saw colors differently from the others. It turned out that the brothers could not see reds and greens accurately. Although Dalton was not a biologist or medical scientist, he became the first person to discover color blindness and the first color blind person to be discovered. He had the vision and mind of a scientist and did not miss out on details, so that he could discover new things. He published his findings and became the first person to study color blindness. Color blindness was later known as Daltonism.

This is an example of the uniqueness of genes. Dalton and his younger brother were colorblind, but his older brother and mother were not.

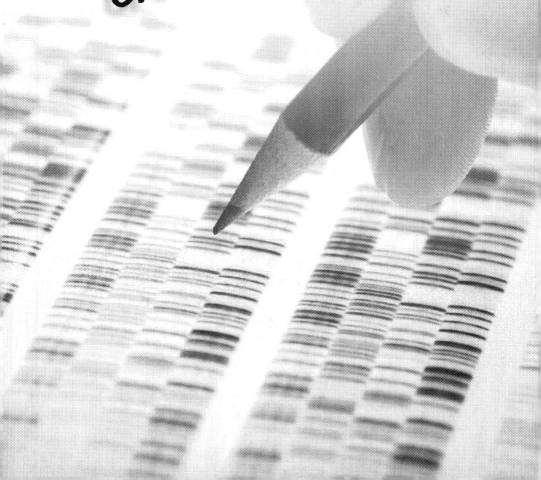

Chapter 3

The Genetic Code

> **" "** Life has amazing characteristics. When you look at your parents, you can find that you have some but not all of their features. How does the genetic code determine your inherited features? What are the characteristics of the genetic code? **" "**

Questions to Ponder

Are there two identical people in the world?

Which of the parents' characteristics will be passed on to their children?

What will happen if a person's DNA is changed?

Why did ancient apes and not other animals evolve into humans?

Genetic medicines can cure certain human diseases.

Can genetic research be used to bring back extinct animals or plants from long ago?

Hint

When reading this chapter, you can think about the questions. You may find answers to them. If you are interested in the questions after reading the chapter, you can search the Internet for knowledge on the specific topic.

Student: Some twins look exactly alike and can't be distinguished from each other. But some twins look completely different and have different personalities. Why is this so?

Lin He: Before we discuss your question, let's see if you can answer my question. There are no two identical leaves in the world. Is this statement correct? Think about it for a moment.

Lin He: Let me ask you another question. Are there two identical people in the world?

Student: I think that, with so many people on the earth, you can find two people alike.

Student: I think there can be two people looking alike because cloning technology is now available.

Student: I think so. I have read about a technology called genetic modification in a book. The technology can transform two people by making their genes the same.

Student: I think there have ever been two people alike. They were just not living in the same time or space. They could have appeared in different periods of time without being recorded in history, so they were not discovered.

Are There Two Identical People in the World?

Genetic Secrets of Twins

Before diving into the answer, let's talk about what determines the sex of the human.

The sex of human beings is determined by the XY system. Each somatic cell contains 22 pairs of autosomal chromosomes and one pair of sex chromosomes. Male sex chromosomes are XY, while female sex chromosomes are XX.

The presence or absence of the Y chromosome determines the male or female sex. The father's sperm cell contains 22 autosomal chromosomes and one sex chromosome, which can be 22 autosomal chromosomes with either X or Y chromosome. The mother's egg cell also contains 22 autosomal chromosomes and one sex chromosome. Because the female sex chromosome is XX, the chromosomes in the egg cell are 22 autosomal chromosomes with X chromosome. With the 22 autosomal chromosomes being the same, the chromosomes that determine the sex are the X and Y chromosomes. In theory, the ratio of boys born to girls or XY:XX is 1:1. Whether a boy or a girl is born is determined by the father's sex chromosome.

Human karyotype

Male

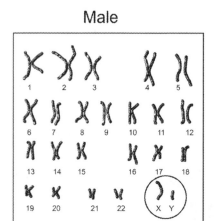

Female

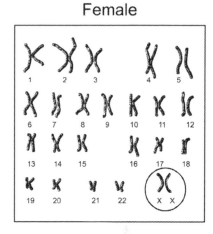

Human somatic cells contain 22 pairs of autosomal chromosomes and one pair of sex chromosomes. Male sex chromosomes are XY, while female sex chromosomes are XX.

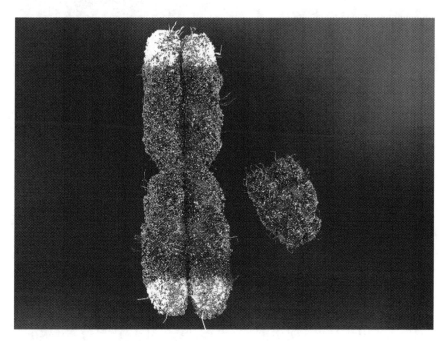

X chromosome (left) and Y chromosome (right)

With twins, there is nothing mysterious about them. In the following section, you can learn about how twins develop.

Humans are advanced mammals that produce offspring through sexual reproduction. We all develop from a fertilized egg.

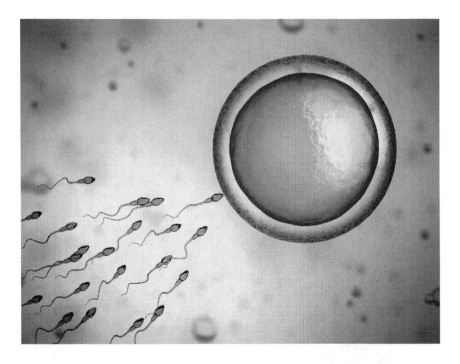

Generally, a fertilized egg develops into a fetus. A fertilized egg however may split into two identical fertilized eggs that develop into two individuals. When two fetuses are from the same fertilized egg, they are called identical twins.

Identical twins are similar not only in appearance but also in blood type, intelligence, even certain physiological characteristics, and in susceptibility to diseases. Conjoined babies sometimes reported in the news are actually identical twins whose organs are joined because of incomplete splitting of the fertilized egg.

Most identical twins are of the same sex, and the pair is unlikely to be a boy and a girl. But in rare exceptions, the fertilized egg loses a copy of the Y chromosome when it divides into two embryos, causing the babies to have XY and X0 sex chromosomes.

An adult woman generally produces one egg every month. Sometimes two eggs are released at the same time for some reason and are fertilized at the same time, producing two different fertilized eggs. The two fertilized eggs each have a placenta and are independent of each other. These babies are called fraternal twins. Fraternal twins look just like ordinary siblings. Twins of the opposite sex are fraternal twins rather than identical twins.

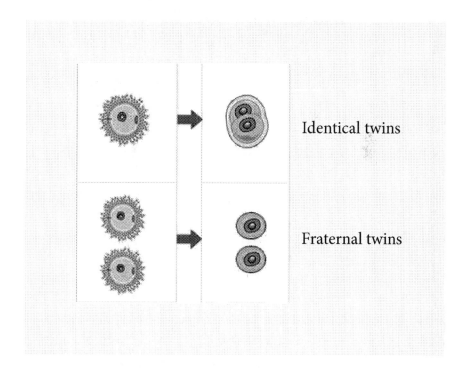

Identical twins

Fraternal twins

You may meet a pair of identical twin brothers or sisters in real life looking so similar that it is quite difficult to distinguish between the two.

Identical twins develop from a fertilized egg that splits to become two identical offspring. The twins have many similar features, such as height, weight, intelligence, fingerprints, and risk of genetic diseases. As identical twins grow, it is possible to find differences between them in terms of physical appearance and personality. There are reasons why differences between identical twins exist:

1. Gene mutation

Errors may occur during DNA replication in cells, causing the DNA to omit certain genes or insert additional gene fragments, and resulting in differences between identical twins. Genetic mutations may also cause one of the twins to suffer from certain genetic diseases like Beckwith–Wiedemann syndrome.

2. DNA methylation

DNA methylation is a process in which the hydrogen atom (H) of some nitrogenous bases in DNA is replaced by a methyl (CH_3) group, changing the function of the gene without changing the DNA sequence. The gene function may change from dominant to recessive, or recessive to dominant. The degree of DNA methylation level is influenced by diet, lifestyle, stress and other acquired factors, which can make identical twins different when growing up. However, methylation is reversible, and the key is to develop healthy lifestyles.

3. Inactivation of X chromosome

The male cell has one X chromosome and one Y chromosome, so one X chromosome is enough for humans. In the female cell, one of the two X chromosomes is inactive, and this phenomenon is called "X-chromosome inactivation" or "lyonization". The inactive X chromosome is called the "Barr body". Because the selection of X chromosome to become the Barr body is random, cells with identical DNA can show different traits. Therefore, there may be great differences between twin girls because of the random inactivation of the X chromosome.

4. Differences in development and growth environment

When the fertilized egg splits into two embryos and the embryos implant in different parts of the placenta, the environment begins to affect the embryos. Differences in blood supply and nutritional supply in different parts of the placenta will result in different levels of development of fetal organs, causing the differences between identical twins. After birth, the growth environment will, for example, influence their personalities and values. Interaction with different kinds of people could also subtly change the twins.

Life's experience shows that no matter how alike identical twins are, their parents will surely be able to distinguish them. Their siblings or close friends can also tell them apart.

Chang Qin: Are there two identical people in the world? One student said that there could be two people, one living during the Tang Dynasty and the other living in the modern era. The only problem is getting them to show up at the same time to compare.

Bo Hai: Professor He, is this possible?

Lin He: This possibility cannot be ruled out. But based on today's knowledge, it is impossible. Why?

Returning to the question "Are there two identical people in the world?", think about your answers again.

Student: Because their DNA has already changed by the time they are born. They may be different even though they are identical twins.

Student: Unlike what he has said about the genes of identical twins being changed before their birth, the twins may be very similar when they are born. But their genetic material changes during the process of development.

Lin He: I think the answers of both of you are quite reasonable. However, you were not able to communicate clearly on the change. What kind of change could it be?

You now probably agree that the genetic material of identical twins is exactly the same in theory. Assuming that their genetic material is exactly the same, they should look exactly the same. But, no matter how similar the twins are, they can still be distinguished by their parents and friends. So, they cannot be exactly the same. There is still the question of how the variations between identical twins come about. Even though twins have the same genetic makeup, differences between them, no matter how small, still exist. Why are there differences?

The human body is a closed open system. Skin and muscles enclose human bodies. The skin also acts as a channel. The skin does not wrap the human body so tightly though. It can allow viruses and bacteria to pass through to infect the body. You may have heard about diseases caused by various viruses or bacteria. Diseases happen because of the communication between humans and the environment. Coming back to identical twins, they should in theory be exactly the same, yet they are different. The major reason for their difference is due to later modification made to the genes.

DNA methylation is a form of DNA chemical modification that can change the genetic performance and control gene expression without changing the DNA sequence. This type of modification is also called an "epigenetic modification".

Let's first understand the relationship among the gene, DNA, chromosome, and cell nucleus. The nucleus is located in the center of the cell. All chromosomes are contained in the nucleus. A single chromosome is composed of DNA in a highly spiraled structure, and a gene is a section of DNA on the DNA strand.

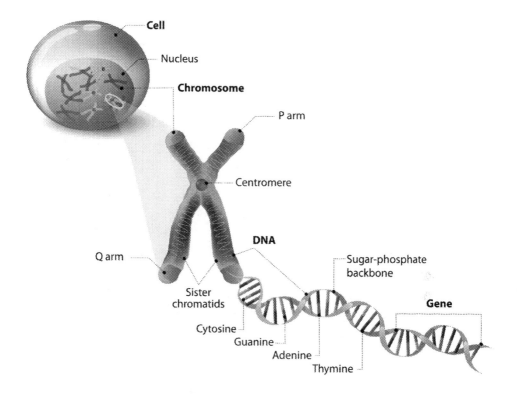

Cell nucleus, chromosome and gene

The environment causes small changes to the genes, resulting in different phenotypes in humans, such as differences in appearance or personality of identical twins. Such differences may arise from modifications to their originally identical DNA by the external environment.

One of the earliest discovered modifications is DNA methylation. DNA is an organic substance. Substances are made up of compound molecules, so is DNA. Methylation is the substitution of a hydrogen atom in the DNA by a methyl molecule. This process is like changing a label in the DNA and is called modification.

The appearance and personality of a person are biologically determined by the genes, and the external state of the person is affected by the growth environment. So assuming that two people living in different times in history are born with identical genes, they will eventually not be exactly the same because their growth environments will be completely different.

Which characteristics are passed on from parents to children?

Student: Professor He, I'd like to ask which characteristics will be passed on from parents to their children.

Lin He: It can be said that all characteristics can be passed on from parents to children.

Bo Hai: But we can only see some of them, right?

Lin He: Yes, in a way. But the random assortment and combination of dominant and recessive genes also need to be considered.

Let's consider an example of the effects of dominant and recessive genes in the offspring.

In the picture, the father's hair is straight and the mother's hair is straight too. But their baby's hair is curly. You may suspect that the baby is not their biological offspring.

Is the baby biologically related to the parents?

There are two types of inheritance: dominant inheritance and recessive inheritance. The sperm and egg combine to form a fertilized egg and they each pass their own genes on to the offspring. You may ask if the father will pass on his small eye trait to his baby. The characteristics of the baby's appearance are mainly influenced by the dominant genes. For example, big eyes are dominant over small eyes; long eyelashes are dominant over short eyelashes; and tongue rolling is dominant over non-tongue rolling. If one of the parents has big eyes, the baby will have a higher possibility of having big eyes. If one of the parents has long eyelashes, the baby is more likely to have long eyelashes. The possibility here is actually probability.

There are alternative forms of a gene called alleles that arise by mutation and are found at the same place (locus) on a chromosome. The gene at a locus can be either dominant or recessive. Dominance is represented by a capital letter and recessiveness by a lowercase letter. In the genetics chart, both parents are able to roll their tongues, and their genotypes are represented by Dd, where D refers to the dominant allele and d refers to the recessive allele. The dominant D allele determines whether you can roll your tongue or not.

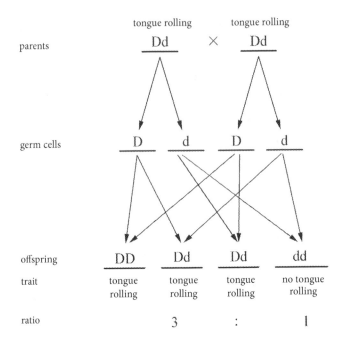

As long as there is the D allele (tongue rolling gene), you have the tongue rolling trait. The tongue rolling genetics chart shows the possible genotypes of the offspring, DD, Dd, and dd, in a ratio of 1:2:1. The first two genotypes will produce the tongue rolling trait, while the third one with the recessive gene will produce the non-tongue rolling trait. Each offspring can only randomly inherit one genotype, based on the probability of each genotype.

The characteristic appearance of the offspring is determined by the dominant allele of the trait. The dominant gene is randomly distributed to the male and female offspring. So whether the baby looks more like the father or mother is also random. The baby randomly inherits the genes responsible for the physical features, such as those of the eyes, nose, mouth and chin, from the parents. The genes from the father and mother exert the same influence. Whether you look more like your father or mother depends on whose genes are stronger.

What Will Happen If a Person's DNA Is Changed?

It is possible to change parts of the DNA by removing existing DNA or inserting replacement DNA using technologies. There is a new technology called "gene editing" that performs like scissors to edit genes with accuracy. If genes were to be edited at the initial stage of the human embryo, the baby born would be different from the original one. It will not be easy to replace the entire genome of a human embryo. Such a practice is a violation of ethics and is not recommended.

DNA editing or gene editing refers to the technology of changing and editing genes by knocking them out or inserting specific DNA sequences. The purpose is to "turn off" certain genes or insert specific genes. Scientists have developed a new gene editing technology called CRISPR-Cas9 (clustered regularly interspaced short palindromic repeats — CRISPR associated system) or CRISPR in short. The CRISPR system was originally used by bacteria to cut out and destroy viral DNA sequences. The system can be used to edit genes of plants and animals, including humans.

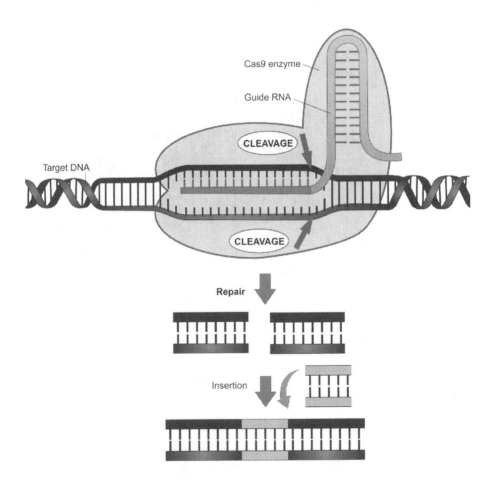

CRISPR-Cas9 system for genome editing

There was a hot topic in the news about a doctor in Italy who proposed to perform the world's first head transplant. What a horrific idea! Once news broke out, the local media began reporting the following:

Valery Spiridonov, a Russian programmer, suffers from spinal muscular atrophy known as Werdnig-Hoffmann disease. This genetic disease has kept him stuck in a wheelchair.

Italian neurosurgeon, Sergio Canavero, has been hoping to perform the world's first human head transplant. Canavero and his colleagues believed that they were able to perform their surgery in 2017. They planned to use electrical impulses to stimulate the nervous system of a donor body, in order to verify the possibility of connecting the spinal cord of the volunteer's head to the spinal cord of the donor body. Spiridonov was so eager to stand up that he volunteered for the surgery.

Canavero's announcement of the head transplant caused a deep concern among scientists. Many doctors participated in related experiments and conducted the experiments on animals such as dogs and monkeys. Chinese doctor Xiaoping Ren also joined the doctors. The scientists thought that more time was needed before determining whether head transplant could be performed in humans. There will definitely be ethical considerations for such a human head transplant.

The question is how you view this news. It is true that the Russian has a serious physical condition. His condition is a genetic disease that is caused by his own genes. Because his condition was incurable by surgery and medication, he put his last hope on transplanting his head to a donor body. Human head transplant is not about changing the DNA. The surgery is not as simple as a liver or a kidney transplant, but involves nerves and consciousness of the person to control another physical body. The Italian doctor had a team to perform the surgery. Based on the current reports, the scientists have performed many experiments on animals, but with a low rate of success. The success rate of this surgery may be extremely low now. But with the accumulation of experimental data and more animal experiments, the success rate in the future may be better. Just like the first liver and kidney transplants, head transplant may one day be successful. Once successful, such a transplant will be accompanied by great ethical challenges.

The latest report says that the Russian has given up on the head transplant. It is not clear if the head transplant will happen in the future.

Can Modifying the Genes of Ordinary People Turn Them into Superman?

Student: Professor He, can we make human beings as powerful as the superheroes in movies by modifying their genes?

Bo Hai: Can ordinary people become supermen by changing their genes?

Lin He: Yes, in theory.

If we can change our DNA, can we also create superman?

You may have noticed that the development of assisted reproductive technology is getting faster and faster. If you visit the fertility hospital, you will find many people seeking assisted reproductive services. I happen to be there a few times, and I could not even get to the elevator from the hospital entrance. It was simply too crowded to move. It is hard to imagine that many people require fertility treatment!

Many couples are unable to bear children by themselves. Their option is to go to the fertility hospital to obtain the eggs or sperms that can be used for *in vitro* fertilization. The *in vitro* fertilized egg is not a naturally fertilized egg and the baby born via this assisted reproductive technology is called a "test tube baby". If the egg from a beautiful mother and the sperm from an intelligent father are used to make a test tube baby, the baby born could be beautiful and intelligent. It is possible to create superman in the same way, but there may be ethical obstacles.

Assisted reproductive technology

Assisted reproductive technology is used to achieve pregnancy in infertile women by medically assisted methods. The two main types of assisted reproductive technologies are artificial insemination, and *in vitro* fertilization–embryo transfer together with its derivative technologies. The test-tube baby is produced from the *in vitro* fertilization–embryo transfer technique.

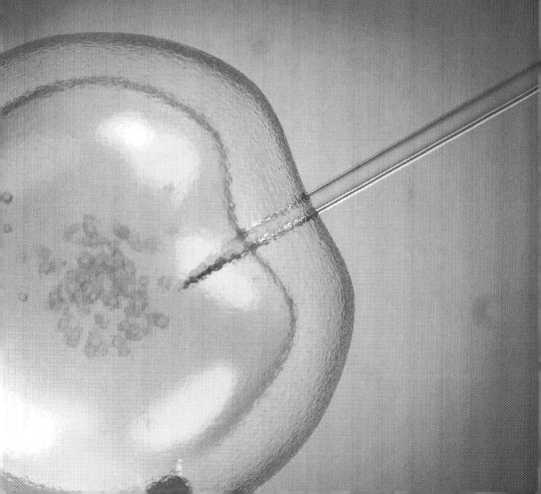

Will there be more and more geniuses?

Student: If genes can be changed, does it mean that, with genetics research, there will be more and more geniuses?

Lin He: People conduct genetic research in the hope of changing mankind to make it better and better. Do you know what genius means?

Will there be more and more geniuses because of gene modification? Before answering this question, find out what people say about geniuses.

Genius is one percent inspiration, ninety-nine percent perspiration.

If hard work matters, why is there genius?

A genius should be born with a high IQ.

I think a genius can see something at a glance.

A born fool can also be called a genius.

I think a genius is very receptive in one area and is good at it.

Some say that a genius is a person born with high IQ. The term "IQ" here is used almost interchangeably with "wisdom", just that they are expressed differently. In fact, geniuses are born and are difficult to made. You probably know Vincent van Gogh. Most people think that van Gogh was an artistic genius, and his artistic achievements were not achieved by hard work alone.

van Gogh was a painter who deeply influenced contemporary painting. He created his unique personal style using vivid colors and powerful expressive techniques. He is also regarded as the pioneer of Fauvism and Expressionism. van Gogh left behind a large collection of works, despite his short life. van Gogh was a genius painter. He did not receive systematic training in painting as a child and only started to paint at around the age of 27. His most well-known works were mostly created in the last two years of his life.

Many articles about the life of van Gogh reveals that he suffered from mental illness later in his life. The painter who expressed life and unrestrained strength using his brush eventually ended his life.

In the eyes of ordinary people, van Gogh's life was full of confusion, especially in the last two years of his life when he was too insane to be understood. His life was gray and violent. He harmed himself and was eccentric, yet his works were so bright, beautiful, and full of life. Such an enormous contrast makes people curious about the talented painter.

In July 1890, van Gogh completed his last oil painting *Wheat Field with Crows*, which is held in the collection of the van Gogh Museum in Amsterdam.

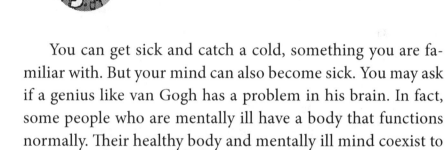

Student: I think a certain part of his brain is different from ordinary people's, perhaps a little more developed.

You can get sick and catch a cold, something you are familiar with. But your mind can also become sick. You may ask if a genius like van Gogh has a problem in his brain. In fact, some people who are mentally ill have a body that functions normally. Their healthy body and mentally ill mind coexist to form a relationship. Although they are mentally ill, they can do the things that others cannot. So back to the question whether geniuses can be nurtured.

The famous movie *A Beautiful Mind* tells the true story of a genius, the mathematician and economist John Nash. Nash was introverted and withdrawn. His teachers thought that his intelligence was below average, and his unconventional approach to problem-solving was not understood by his teachers.

In college, Nash's talent in mathematics was starting to show. At the age of 22, he proposed a foundational concept in his short paper. This concept became known as "Nash equilibrium" in the game theory.

The behavior of this talented mathematician, however, became peculiar. He once showed up at a party on New Year's Eve dressed like a baby. He was also found with long hair covering his shoulders. Nash suffered from auditory and visual hallucinations, which were severe symptoms of schizophrenia. His colleagues had difficulty accepting his behavior, and his wife even divorced him in despair. Despite the divorce, his wife did not give up on him but insisted on taking care of him. She believed that a person with weird behavior can still be accepted by people around him as long as he lives in a relaxed environment. A madman can sometimes be a genius after all.

If you watch *A Beautiful Mind*, you will know a scene in the movie showing Nash's wife giving him pills every day. He then discards all the pills into a drawer. His wife finally opens the drawer and finds it filled with pills. Why did he do that? Have you thought about it?

Student: He probably had some kind of mental illness. His wife gave him pills to prevent the symptoms from flaring up.

Lin He: What does refusing the pills suggest then?

It seems part of the thinking process in a genius goes haywire. If Nash had taken the pills, his thinking would have been suppressed so that he could not have come up with crazy ideas. He could not have expressed his creativity then. Tossing his pills aside, he devoted himself passionately to working on mathematical problems during his mental episodes. So, the connection between mental illness and genius may be closer than you think.

This chapter discusses an important question. It is hoped that research on genes can serve to advance humanity. There has been much progress in genetics, and there is even the technology to modify genes. So, can technology modify humans? Questions remain whether supermen or perfect humans can be created by genetic modification, like those in science fiction movies. What will the consequences be if this is done? Think about such questions and write down your thoughts.

Can a perfect person be created using genetic engineering?

Chapter 4

Can Genes Communicate with the Universe?

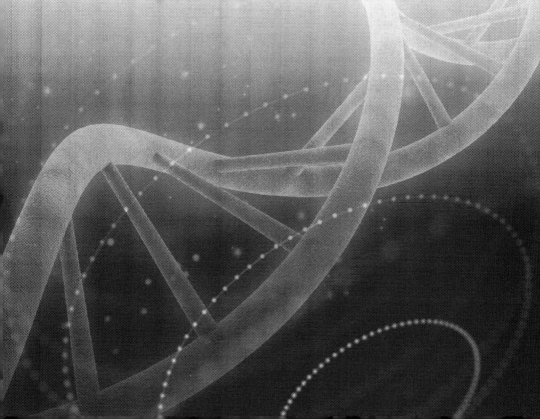

" You are a part of the earth and also a part of the universe, and you are integrated into the universe on this earth. The human body is a closed open system that is continually communicating with its surrounding environment, forming a communication between human genes and the universe. "

Questions to Ponder

What is the relationship between blood type
and genes?

How can the effect of genetic modification
on biodiversity be managed?

Will genetically modified foods affect
the human body?

Hint

There may not be clear answers to many questions, even
after a debate. Put on your thinking cap and come up
with some ideas.

Student: Must children have the same blood type as their parents? Is it possible for parents with common blood types to have offspring with special blood types?

Lin He: There are genetic rules to follow when it comes to blood types. I'm not sure what you mean by "special blood types"? Rare blood types can be found, for example, in people with chimerism. Have you heard of them before? They have two different sets of chromosomes in their cells, which may result in special blood types. In most cases, blood types follow strict rules of inheritance.

Relationship Between Blood Type and Genes

Let's learn about the human blood type. You may hear someone say, "My father has type A blood, and my mother has type B, so I should have...", or "The man with type O blood is a universal blood donor". You may ask where these statements come from and why. Let's start with the definition of blood type.

In the ABO blood group system, the blood type is determined by the antigens on the surface of red blood cells. A person with red blood cells carrying the A antigen has blood type A, those carrying the B antigen have type B, those carrying both A and B have type AB, and those carrying neither antigen have type O. This is a common blood typing system.

Let's learn about the relationship between blood type and genes. The antigens on the surface of red blood cells are determined by your genes. The ABO blood group system is determined by three genes: A gene results in A antigen, B gene results in B antigen, and O gene results in neither antigen.

For ABO blood types, the antigen is controlled by a single gene, which means a particular blood type gene is associated with a particular type of antigen. For example, if a person has A gene, there must be A antigen on the surface of the red blood cells. The blood types of children are related to the genes of their parents. Genes come in pairs, so the six different possible combinations of the ABO genes are AA, AO, BB, BO, AB, and OO. Everyone has only one combination.

In the field of medicine and genetics, the blood types of parents are often used to figure out their offspring's blood types. If both parents are type O, their offspring will be type O. If one parent is type O and the other type B, the offspring can be type B or type O. But if one parent is type A and the other type B, it will be hard to figure out the offspring's blood types because they can be any of the four blood types. Other blood typing tests and techniques can then be used to help in the identification.

Blood Types of Parents and Offspring		
Blood types of parents	Possible blood types of offspring	Blood types of offspring that are not possible
O+O	O	A, B, AB
O+A	A, O	B, AB
O+B	B, O	A, AB
O+AB	A, B	O, AB
A+A	A, O	B, AB
A+B	A, B, AB, O	
A+AB	A, AB, B	O
B+B	B, O	A, AB
B+AB	B, A, AB	O
AB+AB	A, B, AB	O

There are many other human blood groups in addition to ABO, such as Rh, MNS, and Xg. All human blood types are determined by genes on different chromosomes. The gene determining the ABO blood group is located on chromosome 9, while the gene of the Rh blood group is located on chromosome 1. The Rh blood group is classified into Rh positive and Rh negative, and is controlled by two alleles. More than 99% of Chinese and about 85% of Caucasians are Rh positive, while only about 1% of Chinese and about 15% of Caucasians are Rh negative. Hemolytic disease of the newborn is therefore more common in Caucasian babies than in Chinese babies when Rh-negative mothers have Rh-positive babies. The discovery of the Rh blood group is of great importance: It makes the blood transfusion technology become better developed, and the hemolytic disease of the newborn caused by the Rh antigen–antibody reaction can be diagnosed.

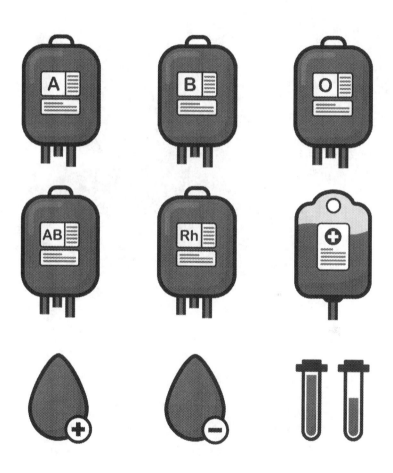

Relationship Between Genetic Modification and Biodiversity

Student: Are there any new genetic drugs being developed to treat certain human diseases? Can we use genetic technology to bring back animals or plants that died out a long time ago?

Student: I wonder if animal DNA can be transferred into the human body.

You have some knowledge of genes, and you are concerned about the current hot topics in society. Let's start with two interesting news reports.

There is an online report on the scientist, Craig Venter.

Making of Man by a "Gene Maverick"

In early 2008, the American journal *Science* published an online article that scientists at J. Craig Venter Institute had not only sequenced a genome but also designed and built the genome from scratch. This means the scientists could synthesize life!

You may wonder who Craig Venter is. Some has called Craig Venter a "gene maverick" because of his bold and crazy approach to genetics research. He has made great progress in the field of gene sequencing and has achieved impressive results for the whole world to notice.

In the mid-1990s, Venter and his team completed the genome sequencing of *Haemophilus influenzae*, the bacteria that cause meningitis. It was the first time all the genes of an organism were sequenced.

In 2000, Venter was ahead of other multinational teams of scientists to complete the sequencing of the human genome. He had boldly predicted that, with the help of supercomputers and his genome sequencing method, his team could complete sequencing the human genome cheaper and faster than the other teams.

In September 2007, Venter published the genetic map of his own personal genetic code in an academic journal, becoming the

first person ever to reveal his own genetic map. On this map, he noted the genes that could increase his risk of Alzheimer's disease, coronary artery disease, obesity, alcoholism, and antisocial behavior. He also has genes for blue eyes which he is proud of.

Venter is as mad as a hatter and can be more than that. He has a crazy mission and dream — to synthesize life!

Scientific research tells us that the DNA of any species is composed of four basic types of nucleotides and that DNA is the basic code of life. Using this code, Venter and his team assembled a new genome for a single-celled organism. Just transplant the genome into a cell and a new species will be born. On May 20, 2010, Venter reported the birth of the first artificial life (cell) in the American journal *Science*. The cell was *Mycoplasma capricolum*, but the genetic material in the cell was artificially synthesized according to the genome of *Mycoplasma mycoides*, and the synthetic cell showed the life characteristics of *Mycoplasma mycoides*. This is the first self-replicating, man-made cell on earth. Venter's team named the artificial cell "Synthia".

Research work by Venter's team to create synthetic cells began as early as 1995. In 2007, Venter's team had already mastered the technique of genome transfer between these two species of mycoplasmas, by using the native DNA from *Mycoplasma mycoides* at that time.

In February 2008, Venter's team successfully synthesized the DNA genome of the single-celled organism (prokaryote), *Mycoplasma genitalium*. The artificial cell "Synthia" that drew world attention was the result of combining these two techniques.

Mycoplasma is the smallest and simplest self-replicating cell. Its genome is the smallest among prokaryotes, making it easy to work with. Although Venter's team had been able to synthesize the genome of *Mycoplasma genitalium*, the organism grows so slowly that the researchers chose the faster-growing *Mycoplasma mycoides* and *Mycoplasma capricolum* as their experimental organisms.

Synthesizing life and new species could benefit mankind or have commercial prospects. It is perhaps really possible to create a multicellular organism that has athletic abilities and is self-aware, with even a face like those terrifying creatures in science fiction movies. If the artificial creature were to get out of hand, the already fragile ecological environment would certainly be threatened. Like in the science fiction novels or movies, life on earth would face a severe test.

Birth of Transgenic Sheep

In early 1998, the Shanghai Institute of Medical Genetics reported that Chinese scientists had created five transgenic goats. The milk of a female goat contained the active, life-saving therapeutic protein, called the human coagulation factor IX, for patients with hemophilia.

This major achievement was the result of hard work by a team of scientists led by Professor Yitao Zeng, an academician of the Chinese Academy of Engineering. The team has brought a ray of hope to the dream of building an "animal pharmaceutical factory" using advanced technology.

"Transgenic" means gene transfer, which refers to the transfer of genes from one organism to another organism. Animals are made transgenic when foreign genes (including human genes) are purposefully introduced into animal cells. After the foreign gene becomes integrated into the animal's own gene, the foreign gene can then proliferate with the division of the animal cells, and be stably transferred to the offspring.

The idea of breeding transgenic animals began in the 1970s. At that time, the genetic medicines were produced, but the cost was too high. The scientists had a bold thought: If the required genes are transferred into livestock with high milk production, such as cow and sheep, the cost of extracting and purifying the therapeutic protein from the milk will be greatly reduced.

The number of transgenic animals has been extremely limited. Since the birth of the first transgenic animal (mouse) in the early 1980s, mass production of transgenic animals has not yet been accomplished. The conception rates from 1989 to 1996 using the zygote microinjection/embryo transfer technology were very low. The Roslin Institute in the United Kingdom obtained 56 transgenic sheep from 2877 samples. The Shanghai Institute of Medical Genetics obtained five transgenic sheep from 119 sheep.

You may ask about the difference between a cloned sheep and a transgenic sheep? To explain simply, the cloned sheep is produced from an unfertilized egg, while the transgenic sheep is produced from a fertilized embryo. The cloned sheep is produced without sexual reproduction, while the transgenic sheep is from the egg produced by sexual reproduction. The cloned sheep is a "copy" of the original animal, while the transgenic sheep obtains a foreign gene and passes it on to the offspring. The value of transgenic sheep is much greater, and the situation will not change anytime soon.

Cloning technology

Cloning is the biotechnological process of creating offspring with identical genomes as the original organisms without the joining of male and female cells. Cloning an organism creates a new living organism with an exact copy of the genetic information as with the original one. A sheep named Dolly was cloned this way.

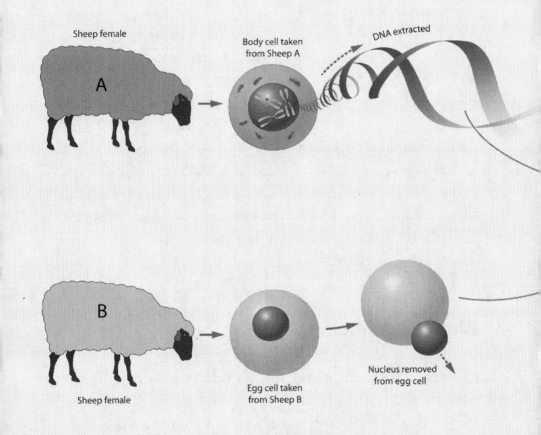

Animal cloning

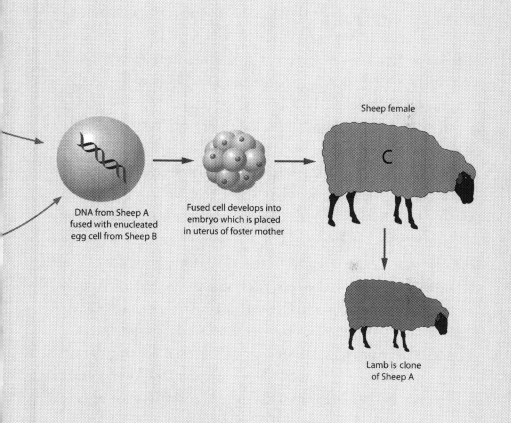

DNA from Sheep A fused with enucleated egg cell from Sheep B

Fused cell develops into embryo which is placed in uterus of foster mother

Sheep female

Lamb is clone of Sheep A

Genetic research has been hot, and the current research focuses mainly on genetic diseases or cancer-related diseases. There is still a long way to go before medicines targeting genes can be tested for use in the treatment of genetic-related diseases.

Scientists are trying to use cloning to bring back certain extinct animal species. Everyone knows that dinosaurs are now extinct. But people want to use the DNA from fossilized dinosaur egg to clone dinosaurs. It is difficult to clone extinct creatures, and no creatures have been brought back to the world before. In the first report, Craig Venter was designing an organism with the help of a supercomputer and making it synthetically. Creating an artificial life has already been done. Nobody knows for sure if genetic technology of the future can bring back the extinct creatures.

You may ask if animal genes can be transferred into humans. The reverse has been performed — human DNA is transplanted into animals. In the second report, the project led by the academician Yitao Zeng from Shanghai is about using transgenic animals to create therapeutic proteins for people with scarcity of certain proteins. For example, if the coagulation factor VIII is to be extracted from the plasma for the treatment of hemophilia A, 12 million blood donors will be needed to get 1.2 million liters of plasma. When the coagulation factor VIII genes are transferred into the mammary glands of sheep or cows, the therapeutic proteins can be extracted from the milk. By breeding transgenic cow or sheep, a stable supply of the protein can be produced in the milk.

Will Eating Genetically Modified Foods Affect Human Health?

Student: Will eating genetically modified foods affect human health?

This is a popular question that causes concern to the society. The issue is very controversial. Why do people have this worry?

Student: The original genes in foods are safe. Once the foreign gene is inserted into the food, the safety of the genetically modified food cannot be confirmed. People may then dislike and even hate genetically modified foods.

Lin He: It would be better if you had thought of another key point: What sort of people would be more worried about genetically modified foods? Why is there a need to genetically modify plants? One reason is to increase production. How can it be done? One method is to use foreign genes. For example, the plant disease resistance gene is inserted into the plants to make them more disease-resistant and eventually more productive. People may ask if the original protein composition will be changed because of the foreign genes that help increase the harvest. Will such a change be harmful to the human body? In other words, will there be any harmful changes in the human body if we eat vegetables carrying such foreign genes? Such concerns are totally understandable.

The DNA of any species is composed of four basic types of nucleotides and DNA is the basic code of life. In fact, all species of living things on earth have the same four DNA bases arranged in different sequences. You can even think that all living creatures belong to one family.

Let's imagine that the different arrangements of these four bases are like family visits. Visiting here and there results in different combinations of the sequence. Genes of plants can be found in humans, while human genes can be found in animals.

It is not known if transgenic plants are harmful and how harmful they are. To date, there is no scientific conclusion on the kind of harm transgenic plants will bring. Although the development of transgenic technology has been rapid, the research in many areas has not been detailed enough, and data have been insufficient to prove certain claims. It is difficult to draw any conclusions on the negative effects of consuming genetically modified foods from either animal studies or human intake.

For the time being, the suggestion is for children who are still undergoing physical development to avoid eating genetically modified foods.

Focus of Debate

You are a part of the earth and also a part of the universe. Living on this earth, you and your genes are integrated into the universe. Your human body is a closed open system, and you are constantly communicating with your surrounding environment and constantly influencing each other.

Many questions have been discussed, but many of them have no answers. Many scientific questions have only possible answers (hypotheses) or get scientists into a friendly scientific debate. This is the charm of science!

Printed in the United States
by Baker & Taylor Publisher Services